An Insider's Guide to Successful Science Fair Projects

By
Felice Gerwitz

Media Angels®, Inc.
Fort Myers, FL

An Insider's Guide to Successful Science Fair Projects

ISBN# 978-1-931941-12-9

Published by Media Angels®, Inc.
Fort Myers, FL 33912
www.MediaAngels.com

Cover photo by Felice Gerwitz
Layout by Christina Gerwitz Moss
Cover by Rebecca Rattner
Editor: Paul Schwarz

Photographs from the DHD Photo Gallery
by Adam Hart-Davis used with permission.
Photos credits: Kathi Kensinger, Carol Benson
Veronique Scanlan, Jackie Perseghetti, and
Felice Gerwitz.

Printed in the United States of America

This book is dedicated to my family.
Thanks for slowing me down!

Anne and Nicholas Gerwitz

Table of Contents

Catch and release ... sand dollars at the beach

Introduction

Science fairs are one of the best opportunities for a student to demonstrate scientific skill and knowledge. It is one of the most overlooked occasions for students because many adults tend to see science fair projects as a burden. I will concede that science projects take a sizable amount of time, yet in my opinion, the reward to the child's self-esteem and character far outweigh the effort.

A science fair project is the culmination of hours of reading, researching, experimenting, investigating, interviewing, writing and ___________ (place your favorite verb in the blank!). Science is multi-sensory; that is, it incorporates all five physical senses. Science also incorporates many subjects, such as reading, spelling, comprehension, listening, mathematics, thinking and logic.

Many adults may feel that it takes a *dedicated* student with a love of learning to *cheerfully* do a science project. While this may not be the case for an assigned topic, which perhaps is only interesting to the adult, a venture into an area of the student's interest will make it difficult to hold him back!

Remember that while science fair projects tend to be family projects, the emphasis should be on helping and directing the student. This is a fine line that I have sometimes crossed with my own children. The project becomes mine, not theirs, when I won't listen to their input!

I hope this book encourages you to seek out information to help your child or student do a project that follows scientific and logical thought. Making a volcano using baking soda and vinegar is not a science project! (Don't feel bad if you've done one; so have we!) I will give you specific guidelines about what makes a great topic and where to find one.

This is by no means a complete or authoritative work on science fairs. Even so, this book will help you understand what is required for entry into upper-level fairs. There are many wonderful books in the library on how to do a science fair project but few that deal with understanding the paperwork that goes on behind the scenes. This work is designed principally to encourage you in working through the entire process of entering a science fair.

You will find a list of recommended resources at the end of this resource. You will also see italicized science keys sprinkled throughout the book. These are helpful hints to remember as you prepare for competition.

One final note: Remember, as parents or educators you are ultimately responsible in helping your child or student to achieve the highest level of success possible. I have heard of countless horror stories relating to science fairs. These situations were the result of a child not adhering to the guidelines required for upper-level competitions, some through no fault of their own. They were misguided, perhaps unintentionally, by those in authority. The results were anything from being disqualified from competition to winning first prize in their school division but being disqualified from upper-level competiton. These situations *may have* been averted by double-checking the guidelines. In some cases paperwork wasn't signed, the wrong (non-approved) materials were used for experimentation, or paperwork was turned in late! I've heard of parents having to hound teachers to get the necessary forms signed and turned in on time. While this seems unnecessary, it happens. Just be aware that parents as well as educators need to be prepared by understanding the rules. I pray that this book helps you to anticipate problems and prevent unnecessary heartache.

Chapter 1

Science Fair Participation and Guidelines

Who may participate in a science fair and are there any guidelines or standards we must follow?

Any student from kindergarten to the 12th grade is eligible to participate in a school science fair. This will vary with your particular school, private school or homeschool affiliation. Only students in sixth grade and above may compete in the regional or state competitions, and only grades nine and above are eligible for the International Science and Engineering Fair (known as ISEF).

This book deals with science fairs in general and higher-level competitions in particular. There are many guidelines that must be followed by students in grades 6-12 who enter a science fair hoping to compete on a regional, state or international level. I have mapped out the basic guidelines in this book, with references to help you in obtaining more information. *These guidelines may change from year to year,* so it is important to receive the correct information from either the coordinator for your school or homeschool group, or the coordinator of the fair you're entering. The guidelines (such as those set by the ISEF—see page 50) may be obtained through your local school board, a fair coordinator or online (www.sciserv.org). I highly recommend you get yourself a copy!

The ISEF has *international guidelines* that are followed by *all of the school affiliates*. These guidelines are the "standard" by which all fairs are measured. In other words, by following these guidelines the chances of reaching upper-level competition increases. If your science fair doesn't follow the ISEF rules, even winning projects will be ineligible for further upper-level competition. Most states follow these guidelines, so the students that win first place in science fair competitions have a chance to participate in the regional, state and—if they are in grades 9-12—international competition.

In order to participate at the regional level you must make sure your school is *an affiliated school.* You can find this out by asking the adult in charge. If you are homeschooling, and plan to hold a science fair, you must register your *school science fair* with your local school board or district, letting them know you are interested in upper-level competition. This can be done by contacting the Secondary Science Administrator or the equivalent for your area. It is important to have one representative for each school or homeschool group as a liaison. I did this for many years, and it made the entire procedure easier. Here in Florida, there's a countywide meeting in early August or September that should be attended by the science representative of our group. If someone cannot attend the meeting for your district, let the district know you are still interested in participating. At the time of the meeting they will hand out the paperwork, requirements and schedules for the upcoming regional science fair.

If your group has not participated in the past, it is important to contact the school district in the spring (or summer) of the year *before* the regional fair. Regional fairs are held in January or February, but quotas are set by the school district in the summer or at the beginning of the school year. The number of students that may be sent varies from school to school. This number is usually based on the total number of students who participated in the science fair in the past. You will be asked how many students are in your group or school. If you homeschool, usually two to three winning students from the homeschool competition will be sent to the regional competition.

If your school or homeschool group does not plan to conduct a science fair, and your child still wants to participate in regional competition, speak to your school board. Perhaps they will help you find a way for your child to compete. I have heard of homeschool students participating in fairs at the local public school they would have attended. The rules in each state vary, so check with the school board for guidelines.

Once you have decided to conduct a science fair, a meeting should be arranged at the beginning of the school year with parents or students, or both. It is very important to begin early! This is a time to go over the rules and hand out all the **necessary paperwork**. In order to compete on a regional level it is necessary to win a school competition in a particular category. (See flow chart on page 61.)

If you are *not* interested in regional competition, the process becomes much simpler; the guidelines may be set by your school affiliation or homeschool group. (See chart on page 60.) A word of caution: Sometimes well-meaning adults wish for the children to complete any project just to say they've competed in a science fair competition. While this is noble, it is wise to teach a child the correct process for completing a science project even at a young age. For example, my little ones have completed projects such as "What Is Color?" They used various watercolor markers to color on coffee-filter paper. When the colors dried, they dripped water on the spots, and *voila!* Some colors broke up into primary colors while others stayed the same. While this wasn't a definitive work and they didn't totally understand the correct scientific concept of color and light, it was a beginning. They learned the correct procedure for completing a science project and were able to complete a project board with my help. They could explain simply what took place, and the groundwork had been laid to bring them to a higher skill level in each ensuing year.

The key to successful science fair projects is to begin early, and obtain a copy of the guidelines and procedures.

Chapter 2

Project Categories

What are the project topic categories?

The basis for science discovery is searching for knowledge in one of the categories below. In the regional, state or ISEF competitions, these categories are the guidelines from which to choose. These descriptions are adapted from the ISEF guidelines. A science project will fall into one of these broad-based categories:

Behavioral and Social Sciences

Behavior of humans and animals, community or social relationships such as: psychology, sociology, anthropology, archaeology, ethnology, linguistics, learning perception, urban difficulties, reading difficulties or concerns, surveys, educational testing issues, etc.

Biochemistry

The chemistry of life processes: molecular biology or genetics, issues dealing with enzymes, photosynthesis, chemistry of the blood, proteins, food chemistry, etc.

Botany

The study of plant life: agriculture, agronomy, horticulture, forestry, taxonomy of plants, plant physiology, plant genetics, hydroponics, algae, etc.

Chemistry

Physical chemistry, organic chemistry, inorganic chemistry, materials, plastics, fuels, pesticides, metallurgy, soil chemistry, etc.

Computer Science

The study and development of computer hardware or soft ware; Internet networking and communications; graphics, simulations and/or virtual reality; data structures; encryption, coding and information theory; etc.

Earth and Space Science

The study of geology, mineralogy, physiography, oceanography, meteorology, climatology, astronomy, speleology, seismology, geography, etc.

Engineering

Technology projects that apply scientific principles to manufacturing and practical uses such as: civil, mechanical, aeronautical, chemical, electrical, photographic, sound, automotive, marine, heat, refrigeration, transportation, environmental engineering, etc.

Environmental Sciences

The study of pollution in air, water, and/or land and ways to control them; ecology

Gerontology

The study of aging in living organisms.

Mathematics

The development of logical systems or various numerical and algebraic computations, and the application of these principles in: calculus, geometry, abstract algebra, number theory, statistics, complex analysis, probability, etc.

Medicine and Health

The study of diseases and health of humans and animals, such as: dentistry, pharmacology, pathology, ophthalmology, nutrition, sanitation, pediatrics, dermatology, allergies, speech, hearing, etc.

Microbiology

The study of biology of microorganisms such as: bacteriology, virology, protozoology, fungi, bacterial genetics, yeast, etc.

Physics

The theories, principles and laws governing energy and the effect on matter such as: solid state, optics, acoustics, particle, nuclear, atomic, plasma, superconductivity, fluid and gas dynamics, thermodynamics, semiconductors, magnetism quantum mechanics, biophysics, etc.

Zoology

The study of animals: animal genetics, ornithology, ichthyology, herpetology, entomology, animal ecology, paleontology, cellular physiology, circadian rhythms, animal husbandry, cytology, histology, animal physiology, invertebrate neurophysiology, studies of invertebrates, etc.

Team Projects

This is a project conducted by two to three students in any of the above categories. (Check with the fair coordinator if you wish to do a team project—the judging criteria changes slightly. See page 46.)

Michael Gerwitz
Science Exploration

Chapter 3

Characteristics of a Winning Topic

What makes a great science fair topic?

A project should contain one or more of the following qualities in order to have the potential of winning:

- A topic relevant to a concern of the day.
- A topic that fits into one of the science and engineering categories.
- A well-researched topic, especially one studied for two to three years or more.
- The development of a better technique.
- The development of a better final project.
- A well-researched and thoughtout project.

The key to a successful science fair project is to focus on new technology, a current issue or a creative approach!

Chapter 4

Science Fair Topics

Where is the best place to search for a topic?

The best place *not* to look is the *library!* The key to finding a topic is very simple. Have your children or students pick a scientific area in which they are interested. Have them think about their experiences. Have they found something interesting growing in the backyard after a rainstorm? Have they noticed an interesting vine or moss growing? Have they wondered about the effects (on their bodies) of glow-in-the-dark watches, or a nearby stereo system? Have they ever questioned why the ink from plastic grocery bags runs when wet, and tried to figure out how to prevent it from running?

After they have explored their experiences, begin looking in the newspaper or in scientific magazines. Keep an ear open for interesting comments people make that may spur an idea. Have the student keep a notebook of science ideas and questions. This ongoing journal will become invaluable as a source of future ideas.

Keep in mind that anything potentially dangerous to the public is prohibited in regional, state and ISEF fairs. No living organisms, including plants or microbes, may be displayed! No animal parts, or photographs containing animals in "other-than-normal" conditions (i.e., surgical techniques, dissection), are allowed, as are human or animal foods, chemicals, etc. Exceptions allowed are teeth, hair, nails, dried animal bones, histological dry-mount sections, and sealed wet-mount tissue slides (ISEF). If your students are interested in doing research involving human subjects, they must obtain approval *before*

experimentation begins (see page 28). Even photographs of people cannot be used without them signing a consent form. Again, it is important to check with the science coordinator, or, better yet, obtain a copy of the ISEF rules for yourself. You can write or obtain a copy online at www.sciserv.org/isef.

How do you narrow down a topic?

Choose one area from the science topic category list and begin to break it down into sub-categories until you have a topic. Let's say, for example, the student has an interest in space. The obvious category to choose is Earth and Space Science. The child may narrow space down to the solar system, then decide he is interested in knowing more about the sun.

Encourage questions, and have the students jot these down. They may wonder about how much sunlight hits the earth at a given time of the day. They may wonder about the shielding effects of sunscreen. They may wonder about solar energy. Once they have a list of ideas, have them highlight the ones that most interest them. Have them form their ideas into questions. Once they have a list of questions, one or two may spur an idea for a science project.

Have the child begin researching what is already known about the topic and narrow it down into a specific scientific problem or question. The goal is to develop an experiment that will solve the problem. Make sure you have discussed the topic with your child or student. Check the guidelines (rules) to see if this project falls within the parameters allowed.

What are some examples of science topics?

If time permits, I would encourage students to make observations before beginning experimentation. The following list is a place to begin your exploration of science topics:

Grades K-3

- How to learn best
- Nutrition
- Organic vs. generic foods
- Food decay
- Food starch
- Soil chemistry, best type of compost
- Gardening
- Study of plants: growing trees, plants, etc.
- Greenhouses
- Seeds, germination
- Computer monitors and eye fatigue
- Natural gas
- Electrical conduction
- Simple machines
- Photography
- Color
- Light
- Pollution in the air, water or land
- Life span of a butterfly
- Insects
- Animals in the wild
- Creatures living in one tree
- Creatures living in one small area
- Bees
- Probability
- Exercise
- Tooth decay
- Eyes
- Yeast, bacteria or fungi
- Magnetism
- Music and plant growth
- Aquariums
- Weather

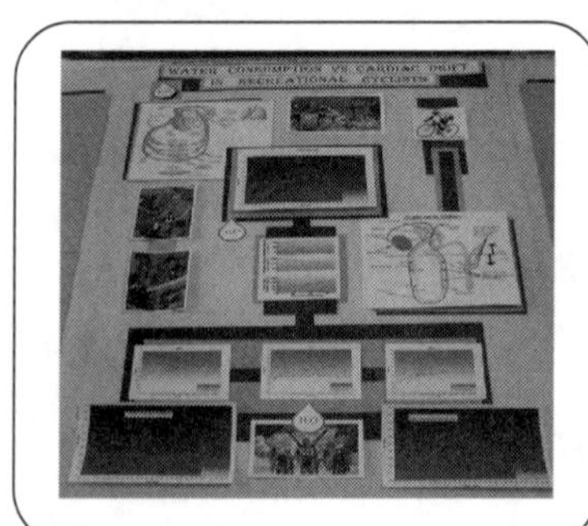

Grades 4-6

- Archaeology studies
- Surveys
- Atmosphere of a home
- Learning issues
- Photosynthesis
- Food chemistry and digestion
- Nutrition
- Organic vs. generic foods
- Vitamins
- Plants, growth techniques and improvements
- Soil chemistry and improvement
- Composting
- Computer basics
- Photography and development
- Geology
- Minerals and the earth
- Value of salt
- Water distillation or filtration
- Minerals and their origins
- Oceans
- Food from the sea
- Shells
- Weather and climate
- Automotive
- Metals and heat
- Heating and refrigeration
- Fire and air
- Transportation
- Fuel
- Pollution
- Erosion
- Earthquakes and other forces of the Earth
- Aging and effects on memory
- Probability

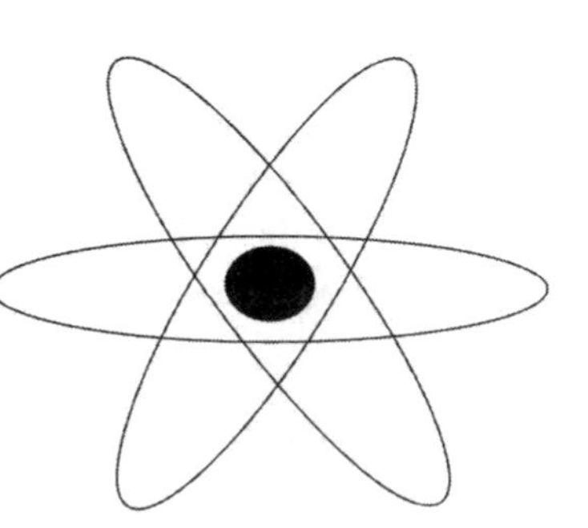

- Teeth
- Speech and hearing
- Fungi
- Yeast
- Microorganisms
- Energy
- Crystal radio
- Acoustics
- Magnetism
- Electromagnetism
- Light
- Mechanics, motors
- Tools
- Paleontology
- Ecology and animals
- Rats
- Invertebrates
- Vertebrates
- Ornithology

Grades 7-8

- Psychology
- Surveys
- Testing
- Germination
- Photosynthesis rates
- Soil and stability
- Plant growth
- Genetic studies within families
- Favorable conditions for fungus growth
- Controlling fungus growth
- Viruses and their effects
- Mold prevention
- Yeast
- Ecological studies of herbicide spraying

- Bacteria in the home, body or soil
- Effects of temperature on elements
- Chemical reactions
- Testing consumer products
- Effects of sunlight
- Effects of salt on rusting rates
- Exercise
- Computer hardware or software
- Internet communications
- Graphics
- Erosion
- Climatology
- Clouds and relationship to weather
- Evaporation rates
- Seismology
- Sound
- Transportation
- Number theory
- Probability
- Accuracy of calculators
- Statistics
- Diseases and prevention
- Ophthalmology
- Sanitation
- Allergies
- Speech and hearing
- Solar energy and design
- Renewable energy sources
- Energy conservation
- Active versus passive solar energy
- Thermodynamics
- Animal genetics
- Ichthyology
- Animal physiology

Grades 9-12

- Human behavior
- Effect of sound on humans or animals
- Effects of hair dye on humans
- Archaeology
- Linguistics
- Learning
- Memory span
- Effects of nicotine
- Nutrition and organic produce
- Genetics
- Hydroponics
- Vegetation and insects
- Sugar and humans
- Weather and humans
- Food chemistry and improvement
- The downside of genetically modified organisms
- Heat and artificial sweeteners
- Soil chemistry
- Soil types and vegetation
- Inorganic chemistry
- Pesticides
- Household products
- Artificial light
- Electrical fields
- Polarized light
- Commercial uses of algae
- Computer simulations
- Virtual reality
- Encryption techniques
- Voices and computing
- Data transfer
- Programming
- New uses of computing
- Mineralogy

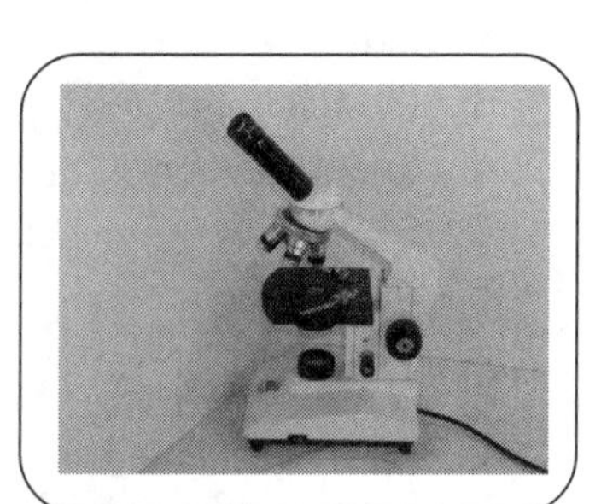

- Erosion control
- Wind-current studies around buildings
- Environmental engineering
- Erosion and shorelines
- Pollution
- Diseases and plants
- Solar cells and energy
- Electrical energy from mechanical sources
- Insulation
- Energy and wind
- Fiber optics
- Fluorescent lighting
- Testing of consumer products
- Pollution and weather
- Mathematics of a snowflake
- Orbits of meteors or comets
- Solar flares
- Photoelectric effect
- Sound dampening
- Techniques of separating elements
- Aging and memory
- Medicine
- Microorganisms and breakdown of petroleum
- Biodegradability
- Decomposition
- Breakdown of oil in salt water
- Laws governing energy
- Entomology
- Conditioned responses of animals
- Wildlife and human interaction

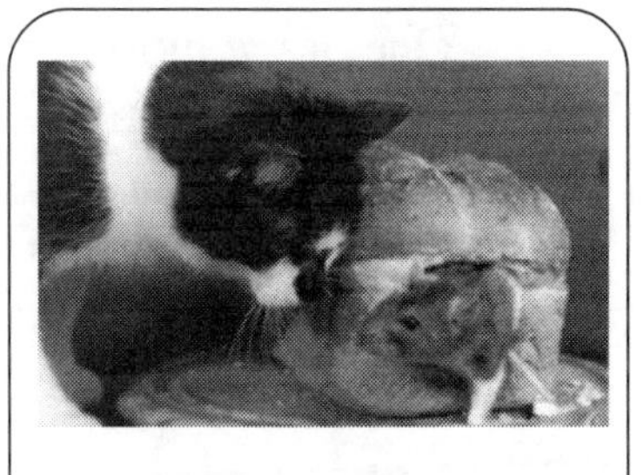

Adam Hart-Davis DHD Photo Gallery

Project Notebook or Journal

How do you keep track of your research?

One of the most important parts of a science project is a journal or project notebook. It is helpful to begin it once a topic is chosen. The notebook or journal should contain all of the information about everything the students are learning and doing on their project. It should be kept in a three-ring folder with pockets for fliers, handy facts, or information that might prove helpful later on when doing the science report.

What is a research plan?

A detailed research plan will be very helpful, as well as necessary for upper-level competition. This plan must be pre-approved before experimentation can take place. It must explain how the student plans to conduct the research and subsequent experiment. It should ask, for example, "What is your question and what do you hope to find?," "What will be needed for the experiment?," and "What type of forms or permission will be necessary before experimentation can take place?"

All this should be written in a project journal. Have the students keep careful records and jot down everything: the names of people interviewed, books, articles, correspondences, Websites, and any other sources of information they've gathered pertaining to the topic. Taking good notes and keeping track of important information will become an invaluable asset and prevent the loss of time spent researching something twice.

Chapter 6

Schedules

To the student...
How to schedule your time:

Once you have a topic, it is important to decide if you have enough time to research and experiment before the deadline for the science fair. Some great ideas take a lot of testing time, so try to begin early.

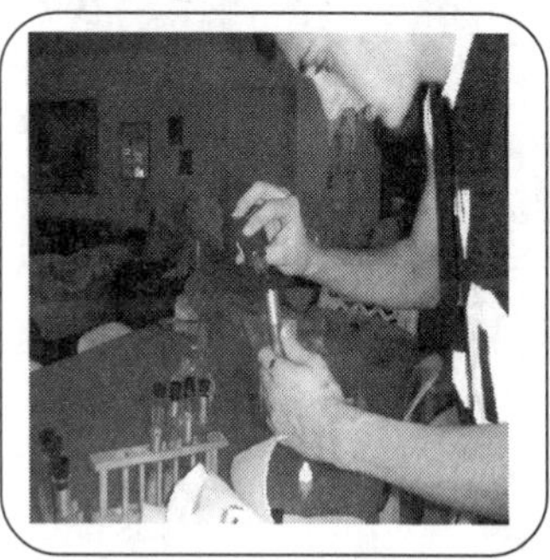

For example, if you have two to three months before the science fair, plan to spend several weeks on research, the major portion on experimentation and the final weeks on preparing your display. It helps to take a blank calendar and fill in the days when you plan to work on your project. Keeping this with your journal or notebook will greatly assist you in time management.

Once you have your question, plan to do exhaustive research on your topic. Don't be surprised if your question changes several times before you make your final decision! As you do your research, you may find your idea has been done many times, or that someone else has already invented what you had in mind.

Don't be discouraged; usually you will be able to think of a different angle. Perhaps a better idea may come to mind as you begin reading the information about your topic.

What if you don't have a great deal of time? (This pertains only to those not interested in upper-level competition.) Sometimes you can't come up with a good idea, or you only have several weeks (yikes!) to

do a project. Don't panic—there are many great ideas out there that can be done quickly. Sometimes it takes being a little creative. I recall a family I know trying to encourage their sick children to cover their mouths when they sneezed or coughed. They never had a problem with this again after conducting the following experiment. The student attempted to grow a culture by coughing and attempting to sneeze (using pepper to aid the sneeezing process) into a bowl of tomato soup. The soup was covered with clear wrap and placed on the counter undisturbed for several days. If my memory serves, there were three bowls: one a cough into the soup, one a sneeze into the soup, and one nothing but plain soup. Guess which two bowls grew amazingly grotesque molds? Needless to say, this project was the topic of conversation at the fair, as well as a winner. It took a few days to research and conduct the actual experiment, and a week or two to create the project board and report.

On the following page I have mapped out a plan for a four-week schedule. Of course, the longer the amount of time you have, the better. Use this schedule loosely and add to this if you have more time. This will give you an idea of how to plan for a science fair and make the most use of your time.

Christopher Scanlan
Photo Credit Veronique Scanlan

Science Fair Project: Student Schedule

To Do List:	Date to Complete:
Week: 1 1. Find a topic. 2. Begin a journal or notebook. 3. Gather information, use a variety of sources: government agencies, professional sources, scientists or businesses, interviews, scientific journals, magazines, newspapers, the Internet, etc.	
Week: 2 1. Read and organize your research. 2. Look at your topic or question and revise it if necessary. Write a project plan. 3. Have the necessary paperwork signed if you are in grades 6-12 and trying to qualify for regional competition. 4. Set up your experiment and create a hypothesis. 5. Purchase or gather your supplies. 6. Keep researching if necessary. 7. Outline your report. 8. Use your resources, for additional help, if needed; doctors, dentists, science teachers, etc.	
Week: 3 1. Experiment! Do your experiment as many times as you are able, pay careful attention to your control. 2. Record your observations in your journal. 3. Begin planning your display board, making a list of necessary supplies to purchase or gather. 4. Jot down ideas for your report and write as much as you can. 5. Take pictures and have them developed and enlarged if necessary (digital pictures are great).	
Week 4: 1. Pray (not kidding :) 2. Complete your report briefly detailing the information you gathered and conclusions you drew from your experiment. 3. Type your report and place text under the appropriate data on your board. Complete your backboard with labels, graphs or charts and visual aids such as clip art or photographs. Keep the board as uncluttered as possible. Display your project notebook or journal and your report with your display board.	

Chapter 7

Paperwork and Forms

What about paperwork and forms to be signed?

(for upper-level competition)

Emily Benson

It is important to check and double-check a student's paperwork if plans are being made for competing outside of the school science fair. One of the ISEF rules requires that an adult supervise the project. This means a parent, teacher, university professor or scientist. The adult is responsible for looking over the topic the student has planned and reading the brief explanation of what will be attempted. Some of the forms must be signed both prior to beginning a project and after completion of the project before competition. The adult sponsor must make sure the experimentation is done in accordance to the rules, oversee the correct completion of the forms to be signed by other adults involved in the review process, understand the criteria for the qualified scientist, and adhere to the guidelines. An adult sponsor and qualified scientist may be the same person. A qualified scientist is defined as one who has earned a doctoral/professional degree in the biomedical sciences, or a master's degree in the area of the student's topic of research. Full details on this procedure are found in the ISEF rules.

How strict are the rules?

The rules are very strict! I have heard many sad stories relating to failure to follow correct procedures in science fairs. One concerned a child who attended a private high school. The child, a freshman, did a science project with mice. She purchased these mice at a local pet store on the recommendation of her teacher. When she won first prize in her school science fair, she learned she was ineligible for regional competition because the mice were not laboratory mice and therefore were outside of the "control norms" required by the upper-level fair. This oversight should have been caught by the scientist overseeing the project, or the teacher in charge of understanding the rules of competition. The student was obviously very disappointed.

Other problems relate to issues with incorrect forms that were submitted. The rules are specific. I know of one student who was rejected from admission to the regional competition because the forms were not signed after experimentation was complete. This oversight could have been prevented. The loosers here are the students, which is why I can not overemphasize the importance of reading the complete guidelines.

Erica Benson
Photo Credit Carol Benson

For upper-level competition forms must be signed both prior to beginning a project and after completion of the project.

Chapter 8

Research

Where is the best place to look for information?

Scientific literature—such as magazines, periodicals and newsletters—can easily be found in college and university libraries or online. A younger child may use the public library, but for higher grades the information for more complex reports is limited. A local college or university library is a better choice.

There are several types of scientific reports, but the two I will discuss are research experimental reports and reviews of scientific literature. These have some of the latest topics and discoveries being discussed. Flip through the magazines, articles and books, and tag anything (my kids love sticky notes!) that catches your eye. Go back later and take a second look. If it still interests you, keep the marker there; if not, remove the marker and continue.

Periodicals are also a good place to look. Don't forget about cross-reference books. You can often look up a topic and find several articles available relating to the topic.

One of the best ways to get information is by letter-writing! Information can be obtained at no (or minimal) charge from departments funded by the government. These addresses are available in the reference section of a college library or online (or see page 72).

Don't forget to tell the recipients that you are a student doing a science fair project and under a deadline to get the information.

Now that the Internet is a household word information is almost instantaneous. Many Internet providers have sites for students containing a section called "homework." Students can post (ask) questions pertaining to various school topics, and teachers will answer them. Under the homework section on one site, my daughter went to the "science" folder and posted a question pertaining to her science project. Within two days she had responses from three teachers and the address of a distributor that carried a chemical she needed to conduct her project. Remember to keep good notes of your sources of information.

The Internet, while holding an abundance of information and ideas, can take longer than you think. Great amounts of time can be wasted looking for information. Use search engines to help you.

What about interviews?

Interviews are a great, often overlooked area of finding science-related information. Many times professional people in the community have a wealth of information they will freely share. Often, science professionals enjoy talking about their craft, with a passion! There are many trade organizations and clubs in many towns. Ours even has an inventors club where retired individuals many doctors and engineers gather to share ideas and mentor others. Look in your local phone book to find resources. Make a list of those within your community who might shed some light on your research, or at least help point you in a direction where you might find another resource.

Chapter 9

The Scientific Method

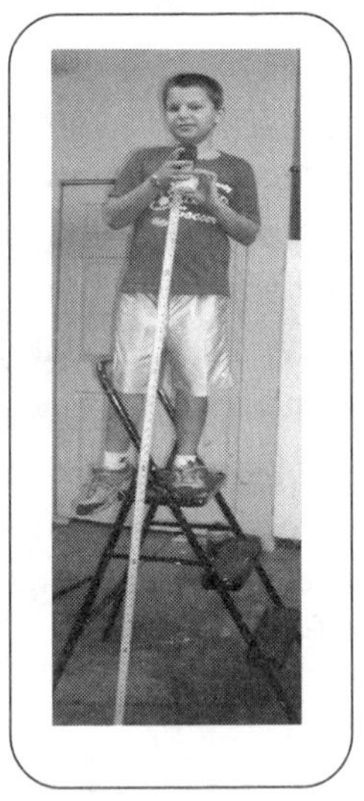

Let's Begin...
What is the scientific method?

The scientific method is the cornerstone of the science fair project. It should be thoroughly understood in order to execute a successful experiment. Warning: Sometimes even science makes unproven claims. This is called faith, not science. Be that as it may, the scientific method is, however flawed, all we have for completing an experiment in an organized fashion. The point of the scientific method is to solve a problem or further study an observation. The steps to the scientific method are: *asking a question, researching, forming an educated guess as to what the conclusion will be (hypothesis), gathering and listing the necessary materials, doing the experiment or executing the procedure, observing the results, organizing the data,* and *stating a conclusion.* Ideally, the conclusion should be the answer to the original question.

Once you have your *question,* you must decide how to find the answer. This will take *research*. Once you have researched your topic, you may begin to form a *hypothesis*. This is an "educated guess" based on your research. You will either prove or disprove this theory or hypothesis. This will take setting up an *experiment*. You will need to figure out what *materials* you will need and the *procedure* you will use to execute the experiment. A procedural plan is a way of testing your hypothesis. In experimentation it is important to use the metric

system because it is used by the scientific community in general. Once you have information from your experiment it is important to correlate this information by comparing two or more *variables*, which may be dependent or independent. A dependent variable can be measured, and an independent variable is one you can control or manipulate.

You also need a control group in order to do an experiment. *A group must have five or more subjects to be statistically valid* (ISEF). A control group is the base used for evaluating the effect of your experiment on the group being tested. A control group is given the same exact treatment as the experimental group, with the exception of whatever you are manipulating.

Here's an example: Suppose you are creating an experiment showing that plants need water to live. You would need separate groups of five to seven plants that are identical in type, size, soil content, position in relation to the sun, etc. One group of plants would receive a full supply of water, one or two groups less water, and one or two groups no water. Change only one variable at a time when experimenting. The experimental groups would be compared with the control group (the plants for which none of the variables are changed and which receive proper care).

By keeping careful records of the experiment (or procedure), you will be able to track any errors you have made, or, if necessary, change your procedure slightly. You then should write your *observation.* Make sure to note if the results prove or disprove your hypothesis. Even if your experiment disproves your hypothesis, it will not count against you in judging, if you can explain why. Either way it is important to do the experiment many times to ensure accuracy and, perhaps, to confirm prior results.

Once you have successfully completed your experiment, it is necessary to organize your data. The organization of data can be in the form of a chart or graph. This adds to the appeal of the display and enables the judges to see your results at a glance. This data could also be in written form which can be placed into a report. The last phase is the all-important *conclusion,* which summarizes your project. It is here that you note if your results differ from your hypothesis, giving a detailed explanation if so.

Chapter 10

The Abstract

Is an abstract really necessary and how do you write one?

An *abstract* is a requirement of an upper-level (grade six or higher) science fair project. An abstract is a brief, clear, concise, 250-word (maximum) summary of your project's purpose, procedure and data, and a short synopsis of your conclusions. If space allows, an abstract should also suggest further extensions of the project, other possible solutions, or more questions to ponder.

You will read in many science fair manuals that the quality of an abstract does not affect the total score. I have found that this does not hold true. Often the abstract is one of the most important parts of the project, especially in regional, state and national competitions. In large competitions the judges, no matter how diligent, do not have much time to spend studying each project. They often read the abstracts ahead of time to get an idea of what the projects entail and perhaps to ponder questions they might wish to ask. Plan to write your abstract as soon as possible, and plan to revise it as you gather more information from your experiment.

One of my daughter's college friends described an experience with a team science project while attending a public high school. Their team submitted a project with which they had competed and won as juniors. Their senior year they continued the project, and it won first place in their category at the high-school and regional levels, yet the team was disqualified from attending the state competition. Among other issues (such as failure of the SRC, to sign their paperwork before

the regional competition), they were told their abstract wasn't good enough. This seems a strange comment since they had already won not one, but two major competitions. Furthermore, this team had been selected to receive a special award at the regional competion given by the Florida Association of Science Teachers! Needless to say, they were very disappointed. Being seniors, they did not have the next year to perfect their abstract or try again! The Scientific Review Committee is also obligated to review every project including abstracts before regional, state and national competitions. They may require unclear abstracts to be rewritten. In this case it simply wasn't done. Teachers have an obligation to help their students by understanding the criteria expected by the ISEF and following them to the letter.

Writing an abstract often takes many drafts. For example, I know several private schools in my area that encourage their participating science students to submit their abstracts to the English teachers. I have personally read contestants' abstracts and suggested improvements. Several of these have gone on to win in upper-level competitions, even though I have a degree in education, not science! I am sure there are many of you who are far more qualified than I. I am familiar with the rules, so when I read an abstract, I expect it to be brief, concise and understandable. Find people in your community who are willing to help in this way. And many people—perhaps some you know—may have science degrees that you are unaware of, so ask around!

Another key to winning a science competition is an abstract! Learn to write one effectively and correctly.

Sample ABSTRACT

used with permission by Kathleen Oare

Project Name: Sulfide Reduction via Marine Chemosynthetic Bacteria: Cleaning Up Oil Spills
By: Kathleen Oare, Copyright 2002.

As a result of petroleum shortages and depletion of oil resources on land in the United States, the oil industry desires to expand offshore drilling in the Gulf of Mexico and is currently investigating the environmental impact. Bioremediation could be used to reduce the concentration of harmful substances such as hydrocarbons and sulfides contained in crude oil in the event of an underwater blowout. The purpose of this research was to determine the effectiveness of *Sulfitobacter mediterraneus* in reducing the amount of sulfide in an aquatic environment.

Twenty-five 10mL samples of sodium sulfide inoculated with 0mL, 1mL, 2mL, 3mL, or 4mL of bacteria solution were tested daily for 3 days for sulfide content. The sodium sulfide without bacteria functioned as a control to ensure the reduced levels of sulfide resulted from *Sulfitobacter mediterraneus* as opposed to oxidation of sodium sulfide. A sulfide test kit was used to analyze each solution to determine parts per million which were converted to molarity.

Statistical analysis conducted via a t-test showed that *Sulfitobacter mediterraneus* significantly reduced the amount of sulfide in the 2mL and 3mL test groups. Although sulfide levels declined in the other groups, the results were not significant. Overpopulation and underpopulation caused the 1mL and 4mL test groups to be less successful. Possible extensions of this research include determining appropriate bacteria-substrate ratios with various sulfide concentrations, investigating the effectiveness of microbes in reducing other toxic compounds, and the application of bioremediation in different environments.

See page 63 for more information on winning science fair abstracts.

Chapter 11

Display Board

How do you set up a display board?

A display board should show all the major parts of your project. In the past, poster board or cardboard was sufficient for grades five and under and if your fair is low-key this may be fine. Many stores now sell free standing science display boards. The typical board size is 30 inches deep, 48 inches wide and 66 inches high. For grades six and above you may use a cardboard or foam backboard, but check with the coordinator. Sometimes it is necessary to have a backboard made of foamboard, wood or other specialty materials. Your backboard should be three-sided. *The dimensions of your board should not exceed 76 centimeters (30 inches) deep, 122 centimeters (48 inches) wide, and 274 centimeters (108 inches) high including the table* (ISEF 2006). These guidelines are very specific for upper level competition.

Your backboard should contain the following headings: *project title, purpose (the question), hypothesis, procedure (experiment), data (which reflects the results), and conclusion (which is the summary).* If this is a two-year project or longer, the title or heading may reflect this and the summary may make brief mention of past research. These headings can be used interchangeably. I've seen boards with the word "question" instead of "purpose," or "conclusion" instead of "summary" it is really up to you.

You can either follow the scientific method in your headings or use the appropriate topics that relate to your specific experiment. As a general rule there is no specific way you must display your information on your board, but it should be in an organized fashion.

Always check the current year's guidelines, though. One year there *was* a specific place required for the abstract. Thankfully I had read the ISEF rules; I watched while other teachers and students scrambled at the regional fair to place their abstracts in the correct position.

Summarize your information in no more than 250 to 300 words. Keep it concise and conclusive—only include information that is important to your project. Don't clutter your board with superfluous facts. Stick to the point! The judges appreciate clear, to-the-point titles and writing. Typed reports are a must in grades six and up. Again, these should be clearly written, making sure to check for grammatical errors.

I know that many parents type the reports for younger (and even older) students. The test here is not on typing proficiency. While it is desirable for the student to complete all the work the content is what is graded, and this should come from the student.

Your display should be eye-catching and informative. A good title is very important. It will get the attention of those walking past. Use stick-on lettering, ready-made titles (sold at office supply stores), or computer-generated banner displays.

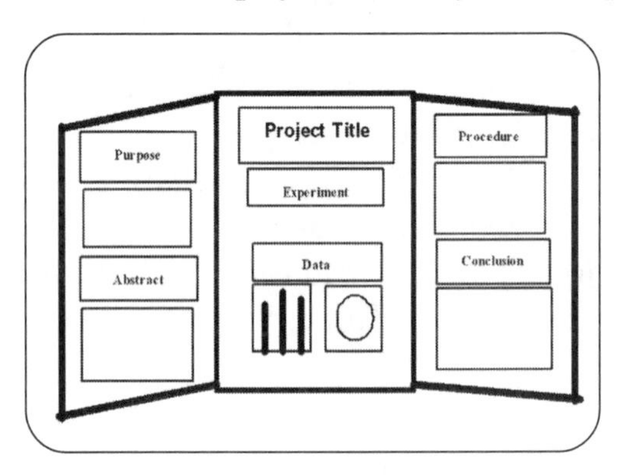

If necessary, use pictures for items you could not bring to the fair. Some photos need signed consent forms, but pictures can improve the display.

Keeping your backboard organized helps the judges read the information you are trying to convey and get a good idea of the results of your experiment. Make the most of the limited space you have, and use color where appropriate. If you are artistic, make use of your talent. The more eye-catching your display is, the better. This should not be your focus, but it is a factor in the grading.

Some fairs do not allow your name or school to be displayed

on the front of the board. Instead you may be given an identification number. Traditionally this has been done to insure the student receives unbiased judging.

Remember these important parts of the display: the abstract, which should be attached to the board and included in the report; the science notebook or journal; and the science report. These should be in three-ring notebooks. In the past we have been advised to ask students to keep extra copies in case anything happens to the original (getting lost or stolen). The science coordinator of our local school district gave this advice at one of the annual meetings I attended. While this is a sad commentary about the world in which we live, it is better to be prepared than sorry.

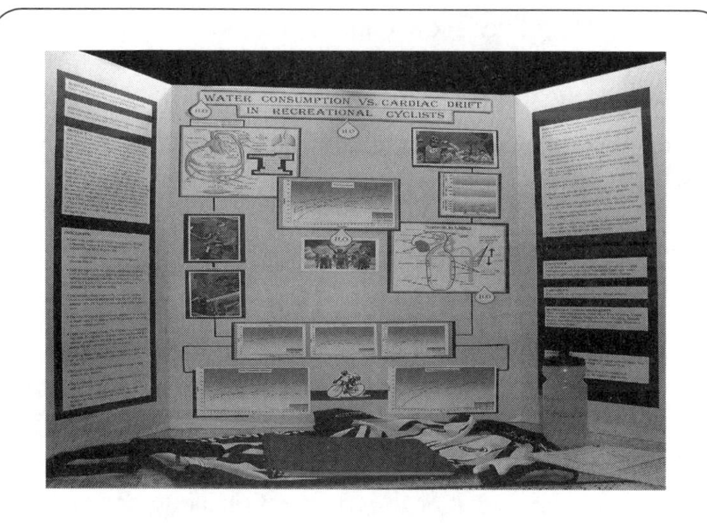

Photo Credit Jackie Persegehetti

The key to a successful project is an eye-catching and informative display board. A good title is important.

Chapter 12

The Report

Why write a report?

The report is evidence of the student's success in researching, planning and completing a project! It should demonstrate scientific originality and creative thinking. A report is a necessary part of science fair projects for grades six and up. Even those in younger grades should attempt a report. This is great practice for those term papers they'll need to learn how to write someday. If they are regularly doing science projects, a term paper will be old hat by the time they are in high school!

To the Student...
What should the report contain?

The report is essentially a formal version of your journal or notebook (see page 24). The report should describe original work (not library research). If you used information from research as the basis for the experimentation, briefly mention the source of the other procedures or experiments, then describe your original contributions.

The report should contain background information about your project, your purpose, your resources, how you acquired information, your experiment and how it was organized, all the data, any diagrams or photos, and your conclusion. You can include detailed information in the form of a diary (which will be easy if you have an up-to-date journal).

If you have obtained information from a library, government

agency, company or organization, it is important to list these resources as references in the bibliography section of your work. Always check the updated requirements.

According to the ISEF rules, your report must contain the following sections: a *title page* containing your name, address, school and grade; a *table of contents;* an *introduction,* which sets the scene, includes your hypothesis, and explains what prompted your research and what you hoped to accomplish; your *method*—the details of your experiment, including the way you collected your data, observations, and any information about design; a *discussion,* the main part of your paper including your results and conclusion, which should flow smoothly from your data and be thorough; a *conclusion,* which briefly and specifically summarizes your results; *acknowledgments*, including credit to those who assisted in the project; and *references/ bibliography,* which should list any documentation that was used in your project.

How specific does a report need to be?

The thought of doing a report does not need to be overwhelming. Keep the language simple for the younger grades and specific for the older grades. Scientists like to see facts laid out in an organized manner, not flowery prose! They are interested in your *results*, your train of thought, what you did, what you would do differently and why you think this is a worthwhile project. The report should be specific enough that someone could recreate the experiment from the information provided. There should be comparisions between the research done and the results of the experimentation. If there were any errors, they should be listed with possible solutions. A well-done report may be submitted to other science contests.

One final note about reports: My niece was part of a high-school science program known as the ISR (Independent Science Research). These students are guaranteed a spot in the regional fair if they complete a science fair project. She placed second in the regional and won an honorable mention at the state level. While she was disappointed not to finish first, she nevertheless had the opportunity to discuss her findings with established members of the scientific community. Upon learning of her desire to enter a prestigious university in an early-admissions program, one of the judges urged her to send her science fair report to the head of the chemistry department. She did, and was delighted to hear back from the university professor. He advised her that he had forwarded a recommendation to the admissions department of this university. This professor wanted her to be admitted based on her 40-page science fair report! (*Way to go, Kathleen! I'm so proud of you.*) Despite not being a first-place winner on the state level, her science report opened doors!

The Presentation

How do you do a presentation?

An important fact to keep in mind when doing the project is that the student may be expected to orally present his research and explain his experiment. This is true in grades six and above, but rarely the case in the younger grades. The oral presentation need not be an area of concern if the student is actively doing the research, experimentation and the preparation of a topic of his choice. He is often more than happy to explain his project to anyone who can stand still long enough to listen.

During a science fair students may have the option of giving a handout to the judges. This is normally limited to the student's abstract in upper-level competitions. In lower-level fairs, the handout rules may be more flexible, so the student may have more freedom to be creative.

To The Student: Presentation advice

You should prepare your presentation well in advance. It is important to practice answering hypothetical questions the judges may ask you. Demonstrate your presentation to family, friends, or any other live audience! Your presentation should contain an explanation of your project and a brief overview of the key points. Do not memorize the entire speech. You will sound more knowledgeable if you are relaxed and not like some automated robot spitting out memorized statistics

or information. Judges enjoy students who can talk freely and enthusiastically about their work. They want to see if you understand your project from beginning to end.

Judges may ask many questions or none at all. Some of the questions they may ask are, "What would have been your next step if you had continued on with the experiment?", "Did you find anything different than you had planned?", or "What didn't you do that you might have if you had more time?"

Another question, "Tell me about your project", needs to be rehearsed. It is easy to answer this direct question, but difficult to know where to begin and how much information to explain.

Remember, the judges have many projects to view in a short amount of time. In order to keep the judges attention, be enthusiastic about your research and findings. Dress neatly and be respectful. This may not be the best time to make a fashion statement. If you are in an upper-level competition it might be wise to bring a book and relax until the time of judging. Best of all, smile and enjoy yourself!

Ben Persegehetti

Photo Credit
Jackie Persegehetti

The key to successful presentation is to be prepared to answer questions, know your project, and smile!

Judging Criteria

What are the exhibition rules?

At the risk of sounding like a scratched CD, it is necessary to remind you to check the current ISEF guidelines. These are often updated from the previous year.

The backboard dimensions given on page 37 may also be changed from year to year, so please keep current. Since science fairs will be under the jurisdiction of the school fair, the person in charge should have all the pertinent information to give you each year.

What do judges look for?

The following is judging criteria adapted from *ISEF Guidelines for Science and Engineering Fairs:*

1. Creative Ability (30 points): How original is the project? Did the project discover a new concept or apply a new technique? Does the creative research support an investigation and help answer a question in an original way? Does the project promote an efficient and accurate method for solving a problem?
2. Scientific Thought & Engineering Goals (30 points): How well researched, understood and thought out is the project? Does it show evidence of using a procedural plan

for obtaining a solution? Was there adequate data to support the conclusions? Does the participant recognize the data's limitations?

3. Thoroughness (15 points): How well researched and thorough (using a variety of resources) is the project?
4. Skill (15 points): How much scientific experimentation (or engineering) was used? What was the level of the experimentation, preparation and treatment of the project? Did the student have the required laboratory, computation, observational and design skills to obtain supporting data?
5. Clarity (10 points): Is the exhibit presented in a way that is easily understood? How clearly is the purpose, procedure, display and conclusions presented? Does the written material reflect the student's understanding of the research? Was the presentation clear-cut, without gimmicks or tricks?

For team projects the point span is a bit different:

1. Creative Ability: 25 points
2. Scientific Thought: 25 points
3. Thoroughness: 12 points
4. Skill: 12 points
5. Clarity: 10 points
6. Teamwork: 16 points

Judges look at how well the student followed the scientific method, the detail that went into the experimentation, and the accuracy of the research. Usually a science fair judge is an expert in his field. For example, a chemist may be a judge for the chemistry division, so he may be familiar with the experiment. The judges also look at whether or not a project is well thought-out, as well as the procedures used for conducting the experiment. They ask themselves, "Did the entrant leave something out, or was the experiment conducted in an accurate and well thought-out manner?"

They also are interested in the significance of your projects.

They may ask, "Is this something of interest to the scientific community, or a problem of the day or some other concern?" They also like clear-cut demonstrations.

The judges may also ask the students specific questions pertaining to their projects, as stated in the previous chapter on presentations. Judges may ask "what if" questions, or what the students would have done differently if they could do the project again. Students should be able to explain their projects and answer questions about their projects when questioned. They are impressed with enthusiasm. They want a student to be able to talk about their report in a knowledgeable fashion.

In my experience, judges are often short on time. With many projects and little time, judges sometimes make a beeline to projects of interest to them, perhaps only giving your child's project a cursory glance. This is the time for polite boldness on the part of your child. Using an opening such as, "Hi, let me tell you a little about my procedure," or "Excuse me, I'd be happy to explain more about my findings," can go a long way in getting a judge's attention, perhaps allowing your child to wow them with the vast amount of knowledge they've obtained!

Who should judge a school science fair?

Judges may range from a community leader with a love for children to a degreed medical doctor. Some good choices include retired science teachers, local college or university professors, science professionals such a chemist or botanist, science teachers from other schools, parents (watch out for conflict of interest, though), or medical professionals from the community.

Remember that judges are volunteers. Try to select those with a love for children and a desire for excellence in education. Whenever possible, supply them well in advance with the science fair date and the judging criteria your school has devised. Make them welcome upon arrival and remember to thank them in writing after the fair is over.

Chapter 15

Conclusion

In conclusion, a science fair can be a wonderful and memorable experience. It is a time when many hours of focused study and research can blossom into lifelong knowledge. Science allows a student the opportunity to find answers to difficult questions and brings about discovery with an investigative approach. It is rarely cut-and-dried. Yet, it can be very exciting because this is a chance for real discovery of new or little-known information.

Understanding what is expected beforehand and making a plan can go a long way toward reducing the stress often associated with this type of activity. I'm not claiming that a science project, or any educational project for that matter, is easy. But once completed, it is rewarding to see the finished product, whether a student wins the competition or not.

Although, winning can be rewarding! Winners in school science contests may receive prizes such as trophies, T-shirts and certificates. In the regional and state fairs, prizes range from awards and special citiations to gifts and cash prizes. Prizes may range from $100 to $50,000 (at the time of this writing) in upper-level competitions!

Projects are a way of life, and not often can you escape them. As a college freshman, my daughter completed two detailed projects in two different subjects! She sailed through both, writing reports, making display boards, color-coordinating the backgrounds, and making banners with letters printed on a color printer. Her confidence came from years at home making similar project boards for her science competitions. (Incidentally, my daughter said the handful of projects that really stood out were from those who had participated in science fairs in the past! How did she know this? She asked!)

Even with the huge array of resources and availability of materials to see you through, there is still only one thing that can make this anything but enjoyable—you! So gather up your enthusiasm and dive into that project! You know it will be worth it . . . once it's done. Happy experimenting!

School, Regional, and State Science Fair Winners

ISEF Guidelines

(Source: International Rules for Precollege Science Research: Guidelines for Science and Engineering Fairs)
Remember, these guidelines may change. Check for updates.

All science fairs serious about the potential to send winners (grades 6-12) to regional, state or (grades 9-12 only) international competition follow these rules. I've added brief comments in small print to help explain some of the details in the guidelines. Check the glossary for further detailed definitions on page 53.

Requirements:

Every student completes Research Plan (1A) and Approval Form (1B) and has it reviewed and signed by an Adult Sponsor. The Adult Sponsor completes the checklist for Adult Sponsor (1).

Certain projects require very specific forms. Experiments dealing with nonhuman vertebrate animals (no pain or discomfort allowed), pathogenic agents, controlled substances, recombinant DNA, or human/animal tissue require prior approval from the Institutional Review Board (IRB) (all federally registered research institutions have IRBs) or Scientific Review Committee (SRC) before experimentation begins (item #8, Research Plan 1A). (An SRC can be made up of a minimum of three members: a biomedical scientist [i.e.: Ph.D., M.D., D.V.M., D.D.S., D.O.], a science teacher (or parent is acceptable) and one other member [school administrator, psychologist, medical doctor or registered nurse, for example].)

All projects involving nonhuman vertebrate animals, pathogenic or potentially pathogenic agents, controlled substances, nonexempt recombinant DNA, certain tissue studies and all studies involving more than a minimal risk in human subjects must have a Qualified Scientist overseeing the project.

All projects involving human subjects must be approved by a properly constituted IRB before experimentation begins.

Informed Consent Form (4B) is required for all subjects in projects involving more than a minimal risk and is recommended for all projects involving human subjects. A copy of any test, survey, or questionnaire must be provided for parental review for subjects under 18 years of age.

Any proposed changes in the Research Plan (1A) and Research Plan Attachment by the student after the initial approval must be approved by the same committees (SRC), before experimentation continues.

Each student or team must submit a maximum of 250-word, one-page abstract which summarizes their work.

Each student should display a project data book and research paper.

All signed forms, certificates, and permits (keep in notebook) must be available for review by an SRC before each fair student enters to compete in an upper level fair. (At the international level the SRC committee is on the premise and will review the materials before entering.) This form must be signed before competition.

Projects which are continuations of last year's work and which require approval (SRC) must be reapproved prior to experimentation/ data collection for the current year.

Any continuing project must document additional research to show it is new and different. (Continuation of Projects Form 7)

If any experiments are conducted in an institutional or industrial setting any time during the project year, a Registered Research Institutional/ Industrial Setting Form (1C) must be completed.

Projects must adhere to all Federal, State or local laws and regulations.

All projects must adhere to the Ethics statement. (Scientific fraud and misconduct is not condoned at any level of research or competition. Plagiarism or presentation of other researchers' work as one's own—and fabrication or falsification of data will not be tolerated. These projects will be disqualified.)

Caroline Kensinger
Photo Credit Kathi Kensinger

Limitations

(ISEF Guidelines continued)

Individual/Team Projects:

Students may only enter one project which covers research for one (12 months) year. Team projects may have a maximum number of 3 students.

The exhibits must adhere to the safety and size requirements (see specific rules).

Continuation Projects:

Students will be judged only on the current year's project. The display may mention the years in which the student has worked on the project. (For example, this is a Two-Year study.) Supporting data from the previous years study (not research papers) may be exhibited on the table if properly labeled.

The continuation project must demonstrate new and different research. (There are specific forms and procedures that must be adhered to for these types of projects. Check current years rules.)

Team Projects:

Team projects compete against other team projects in a separate category. Teams may have up to three members. A team project can not change once experimentation begins. A team may continue with two original members if the third member drops out, but may not take on a new member.

Each team should appoint a leader to coordinate the work and act as a spokesman. Each member must be able to take on this position if necessary. The final work should reflect the efforts of all members.

There are specific forms for team members to fill out, check the rules. The full names of all the team members must appear on the abstract and forms.

Glossary of Terms

Abstract: A maximum 250-word synopsis of the project after it is completed. It should contain the purpose of the experiment, the procedures used, the data collected and the conclusion.

Adult Supervisor: An adult who is familiar with the specific procedures used in the investigation and is approved by the qualified scientist. This can be the teacher or parent who supervises the project, and provides education and guidance during the entire research and experimentation process.

Common laboratory animal: an animal bred for the specific purpose of being used in scientific experimentation. This includes rats, mice, hamsters, gerbils, guinea pigs and rabbits.

Continuation project: This is a second-year, third-year or longer project. The continuation project may reference past work, but must follow a new line of investigation.

Hypothesis: An educated guess as to the answer to the scientific question which is the basis of the project. This guess should be made after thorough research.

Informed Consent: A signed agreement by a human subject, parent or guardian acknowledging that the proposed research and any possible risks have been defined and understood before participation in an investigation.

Institutional Laboratory: An established laboratory within an academic, commercial, medical or government setting, but not in a home or high school.

Institutional Review Board (IRB): A committee of qualified individuals within an academic, commercial, medical or government setting, but not in the home or high school.

Paperwork: Double check and fill out correctly to keep from having a project disqualified.

Qualified Scientist: A person who has a doctoral degree in science or medicine, or a master's degree with equivalent experience and/or expertise, and who has a working knowledge of the techniques outlined by the student in the research plan.

Report: The final written paper detailing the results of the research, hypothesis, rationale of the project, procedures, data, observation and conclusion.

Research: The information gathered in the process of doing an experiment. This includes literature read prior to conducting the actual experiment.

Research Plan: A required written paper outlining the hypothesis and experimental rationale; proposed procedures which the student is planning to use during the research; a description of the methods, techniques and supplies/materials; and the amounts to be used. Sources for acquiring the supplies and materials must be listed.

Scientific Method: The procedures used by the scientific community to conduct experiments in an organized fashion. It consists of the question, the research, the hypothesis, materials used, the procedure, experimentation, observation and organization of the data, and the conclusion.

Scientific Review Committee (SRC): A group of individuals responsible for the evaluation of student research, the research plan, exhibit and compliance with the ISEF guidelines, and rules.

Science Journal

Sketch

Science Journal

Sketch

Date	Source/ Research

Checklist: K-5 Science Fair Project

	Use the current guidelines set up by your school or organization. Be aware of restrictions and regulations regarding the use of animals, human subjects, or hazardous items. Check with an adult before experimentation.
	Look around you. Are there any questions you might have that are science related? Brainstorm with your parents or your friends. Select and narrow down your topic in a specific science category and identify your question. What is your hypothesis?
	Check the amount of time you have to do a project. Make a plan and use a calendar to write out a schedule.
	Begin a journal or project notebook listing all important information as you begin your research; list all resources, procedures, experiments, findings, and people you spoke with, etc.
	Write and send letters, to organizations or agencies that may have necessary information. Use the Internet (with adult supervision), if you wish to obtain information quickly via research or email.
	Organize your procedure—the way you will test your hypothesis. Set up experiments. Do as many experiments as possible—be thorough!
	Organize and record all of your observations and data. If possible, display your data with some type of graph or chart.
	Organize your backboard, make sure to include all necessary categories. Make your display as attractive as possible. Add diagrams or photographs of your findings if you can not bring the actual object to the fair.
	Be ready to explain your project to your family, your friends, and finally the judges. They may ask you questions such as, "What would you do differently."

Checklist: 6-12 Science Fair Project

	Use the current guidelines and regulations for regional, state or international fairs. Be aware of restrictions and regulations regarding the use of animals, human subjects or hazardous items. Check with an adult before experimentation and have necessary paperwork signed.
	Select and narrow down your topic in a specific science category, then identify your question. What is your hypothesis?
	Check the amount of time available to do a project. Make a plan, then use a calendar to write out a schedule.
	Begin a journal or project notebook listing all the important information as you begin your research; list all resources, procedures, experiments, findings, people you spoke with, etc. This is very important!
	Write all letters or emails, sending copies to organizations or agencies that may have necessary information. Use the Internet if you wish to obtain information quickly.
	Write out an experiment plan and have it approved by a scientist (check guidelines). Make sure you have all of the proper forms signed before you begin experimenting!
	Organize your procedure—this is how you will test your hypothesis. Set up experiments. Do as many experiments as possible—be thorough! Keep accurate records.
	Organize and record all of your observations and data. If possible, display your data with some type of graph or chart.
	Organize your backboard, making sure to include all necessary categories. Make your display as attractive as possible. Add diagrams or photographs of your findings if you cannot bring the actual object to the fair.
	Remember to check your forms and have any papers signed before you enter each fair (especially regional level and beyond).
	Have your project journal or notebook ready. Write your report, and a 250-word maximum abstract. Be prepared to present your project orally and answer any questions the judges may ask you.

Setting Up a School Science Fair

Not for upper-level competition

1. Decide on a science fair coordinator. This person will be in charge of setting up the competition and devising the rules. This can also be a committee of just one person.

2. Decide on guidelines for judging criteria and on dates for the competition.

3. Schedule your school science fair. Either late October or the first week of December is best, but you can decide on another time.

4. Plan a meeting with students and parents. Go over all of the necessary requirements. If you cannot meet with both, send home a detailed letter describing what is required of science fair participants.

5. Review the students science topics before they begin experimenting. Give suggestions and advice if necessary.

6. Have students begin their projects.

7. Order ribbons, certificates, trophies, etc.

8. Set up the actual science fair. Contact judges, checking for conflicts of interest. Use local politicians, science teachers from other organizations or schools, retired science teachers, college professors, etc.

9. Check on the student's progress. Ask for project names for inclusion in the science fair program (if one is being created).

10. Check on details for the science fair. Make sure judges have confirmed their availability.

11. Have the science fair!

12. Choose winners to be presented with first, second and third-place awards (and honorable mentions if you desire).

Notes:

Setting Up a School Science Fair

for Regional, State or International Competition

1. Become familiar with the ISEF rules. Obtain a copy.

2. Make sure your school or organization is an affiliated school. Sign up your school with the local school district and make sure your district is ISEF affiliated. Check all dates for upper level competition.

3. Plan to attend any district meetings and gather information to bring back to your group. Set up a SRC committee.

5. Schedule your school science fair. Make sure it is between the later part of October and the first week of December.

4. Plan a meeting with students and parents. Go over all of the necessary paperwork. If you can not meet with both, send home a detailed letter describing what is required of science fair participants.

5. Review the students' science fair plan and have it <u>signed</u> by the appropriate people (SRC/IRB) <u>before</u> the student begins experimenting.

6. Have students begin their projects.

7. Order ribbons or certificates

8. Set up the actual science fair. Contact judges, check for a conflict of interest. Use local politicians, science teachers from other organizations or schools, retired science teachers, college professors, etc.

9. Check on the students' progress. Ask for project names for inclusion in the science fair program (if they are being created for this occasion).

10. Check on details for the science fair. Make sure judges have confirmed their availability.

11. Make sure all of the paperwork is signed by the SRC committee once the students have completed their projects. Check the ISEF rules for the exact information on who makes up a SRC committee and when the forms need to be signed.

12. Have the science fair!

13. Choose winners which will attend the regional science fair based on your quota. Check all forms and submit to the school district for inclusion to the regional fair by the deadline.

14. Check regional fair date. Have winning students review their projects and revise their abstracts or display boards as per rules.

Flow Chart for Science Fair...

Start Early! Begin planning the science fair at the end of the previous year to allow students time to research different ideas during the summer.

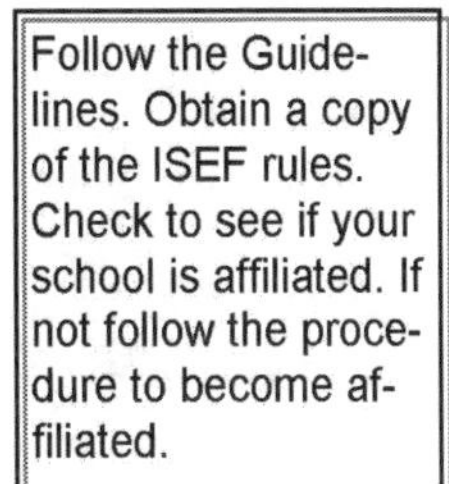

Follow the Guidelines. Obtain a copy of the ISEF rules. Check to see if your school is affiliated. If not follow the procedure to become affiliated.

Prepare timeline of **Key Dates**: For example: October-December: School science fair; January-March: Regional; April: State Fair; May: International Science Fair.

Attend workshops for parents and teachers. Students should also attend informational workshops.

Direct students to select a topic and begin research. Teacher approves projects and reads the student's research plan.

Student submits research plan for prior approval and must have all paperwork signed by SRC/IRB committees before beginning experimentation (see rules).

Explain judging and evaluation criteria.

Students continue their projects, researching, experimenting, recording observations, writing the project report, abstract and completing their backboard display.

Students complete their projects and practice presentations before a peer group, parents, or other adults.

Students compete in a school science fair. Some winners will be chosen to compete in the regional science fair and beyond!

Writing an Abstract

To the student:

An abstract is one of the most important parts of a well done project. Here are the details for writing a clear, precise abstract, consisting of no more than 250-words. Some fairs require that a bibliography be included in your abstract. Several copies must be made and one posted on your board. Check the rules!

1. The purpose of the experiment: Write an introduction statement explaining the main reason you are conducting research into this topic. Write a statement about the problem and the hypothesis.

2. The procedure used to conduct the experiment: Write a summary of the key-points and a brief overview of how you plan to execute this. Include only work you have done. There is no need to include materials unless it influenced the procedure.

3. Data: This should contain key-results that point to the conclusion you have found. Due to word limit there is no need for extensive details or graphs, etc. Keep it brief and specific.

4. Conclusion: Briefly describe your results which should tie into your purpose and hypothesis. Summarize your process. If words permit, include applications of your findings or further investigation possibilities.

Samples of Winning Abstracts

Project Name: The Effect of Microwaved Food on the Weight of Mice: Zoology
Used with Permission by Gabriela T. Martinez, Copyright 2002.

The reason for this project was to determine whether the consumption of microwaved food would have any effect on the weight of mice.

Ten white laboratory mice, two tanks, and other supplies needed for the comfort of the mice were purchased. The tanks were set up in a suitable location. A water bottle and sufficient bedding was placed in each tank. Five mice were placed in each tank; one group of five was to be the control group and the other group was to be the experimentals. A small scale, to be used to weigh the mice, was placed in between the two tanks.

The experiment consisted of weighing the mice every day before feeding them. By recording the daily weights, the progress of the mice could be charted. After the mice were weighed, they were fed. Ten pieces of food, that had been first placed in a cup full of ordinary tap water for five minutes and then microwaved for six minutes, were placed in the experimental tanks. Ten pieces of regular food, soaked for five minutes in the tap water, were placed in the normal tank. This concluded each day's work.

The result of this experiment was that the normal mice gained weight faster than the experimental mice did. The experiment also showed that the final weight after three weeks was greater in the normal mice than the experimental mice. This shows that there may be a deleterious effect in microwaving food.

Project Name: Reinventing the Modern Wheel
Team Project, Luke Carlson, Sarah Cooke and Kristina Cotrell
Used with Permission by Luke Carlson, Copyright 2002.

Last year a low-speed, low head waterwheel was determined to be the solution to an archeological mystery left by the Romans at Barbegal, France. This year, that ancient design was tested as an alternative energy source to meet modern-day energy needs in areas with similar geographical conditions.

Using a bicycle wheel as its core, a wooden waterwheel two meters in diameter, a mounting system, and breastworks were constructed. Water flowed from a small pool through a sluice constructed with 4" PVC pipes into a wooden reservoir attached to the wheel's mounting system. The waterwheel was placed downhill on a shallow slope, allowing approximately a one meter head. An alternator, a pulley system used to increase rotational velocity (RPM), a lightbulb, and an ammeter-voltmeter (used to measure output) were attached to the system. Output averaging 14.7953 watts (P=IV) was obtained from this setup.

This energy output was less than expected, and the reservoir-sluice was raised to approximately a 1.33 meter head, which eliminated the resistance of backflow, and added to the potential energy of the falling water. This setup generated 28.0498 watts (average), enough power to enable the alternator to energize itself and still produce usable power.

These results show that this design of waterwheel generates enough power to provide a viable alternative to non-renewable energy sources. This waterwheel system could be applied in geographical areas with a limited flow and low head of water, and compares quite favorably with known energy yields of commercially available alternative energy sources.

Judging Criteria
K-5

Criteria:	Possible Points: Circle One	Points Awarded
Creative Ability: Original idea	1 2 3 4 5	
Scientific Thought and Engineering Goals: Appropriate problem, follow-through, good conclusion from data	1 2 3 4 5	
Thoroughness: Well researched and thought out, extensive research.	1 2 3 4 5	
Skill: Knowledge, oral presentation well done	1 2 3 4 5	
Clarity: Clear-cut project and exhibit.	1 2 3 4 5	
Additional Notes:		Total Score:

Judging Criteria

6-12

Criteria:	Total Possible Points	Points Awarded
Creative Ability: How original is the project? Did the project discover a new concept or apply a new technique? Was an efficient and accurate method used for solving the problem?	30	
Scientific Thought and Engineering Goals: How well researched, understood and thought out is the project? Is there adequate data to support the conclusions?	30	
Thoroughness: How well researched and thought out is the project? Were there a variety of resources used for the project?	15	
Skill: How much scientific experimentation (or engineering) was used? What was the level of preparation, experimentation and treatment of the project?	15	
Clarity: Is the exhibit presented in a way that is easily understood and clear-cut? How clearly is the purpose, procedure, display and conclusions presented?	10	
Additional Notes:		Total Score:

Judging Criteria

Team Projects

Criteria: **For Team Projects**	Total Possible Points	Points Awarded
Creative Ability: How original is the project? Did the project discover a new concept or apply a new technique? Was an efficient and accurate method used for solving the problem?	25	
Scientific Thought and Engineering Goals: How well researched, understood and thought out is the project? Is there adequate data to support the conclusions?	25	
Thoroughness: How well researched and thought out is the project? Were there a variety of resources used for the project?	12	
Skill: How much scientific experimentation (or engineering) was used? What was the level of preparation, experimentation and treatment of the project?	12	
Clarity: Is the exhibit presented in a way that is easily understood and clear-cut? How clearly is the purpose, procedure, display and conclusions presented?	10	
Teamwork: How well did the group work together? Did each member understand the project?	16	

TOTAL SCORE:

Evaluation Worksheet

Student (s) Name:
Name of Project:
Category Entered:
Judge's Name:
Date:

Category	**Possible Points**	**Points Earned**
Creative Ability		
Scientific Thought & Engineering Goals		
Thoroughness		
Skill		
Clarity		
Dramatic Value		
Additional Notes:		
Totals		

Science Related Websites

International Science and Engineering Fair (ISEF)	sciserv.org/isef
American Association of Physics Teachers	aapt.org
National Academy of Science	pnas.org
Mathematical Association of America	maa.org
National Academy of Sciences	nas.edu
NASA Spacelink	NASA.gov
National Council of Teachers of Mathematics	nctm.org
National Science Foundation	nsf.gov
National Science Teachers Association	nsta.org
Technology of Americans	iteaconnect.org
U.S. Department of Education	ed.gov
National Science Council	nsc.gov
Science (Information provided by US government)	science.gov

Contest Information

International Science and Engineering Fair (Grades 9-12)
Science Service Inc.
1719 N. Street NW, Washington, DC 20036
(202) 785-2255
Web site: www.sciserv.org

Intel Science Talent Search
c/o Science Service
1719 N. St. N.W.
Washington, DC 20036
Web site: www.sciserv.org/sts

Discovery Channel Young Scientist Challenge
(contest for grades 5-8) Web site: www.discovery.com/dcysc

Additional Resources

Science Fair Projects Online:
www.science-fair-projects-online.com
www.scienceproject.com (for-pay site)
www.educationplanet.com
www.scifair.org
www.homeworkspot.com (ask questions)
www.goenc.com
http://education.usgs.gov
www.imagineeringezine.com/e-zine/science.html
www.MediaAngels.com (science project articles and information)

Online Libraries
Internet Public Library: www.ipl.org
Library Spot: www.libraryspot.com

Reference Books

1001 Ideas for Science Projects; Marion A. Brisk; MacMillan Publishing Company; 1998.

Blue Ribbon Science Fair Projects; Vecchione, Glenn; Sterling, 2005.

(The) Complete Handbook of Science Fair Projects; Bochinski, Julianne Blair; Jossey-Bass; 2003.

How to Do a Science Fair Project; Tocci, Salvatore; Franklin Watts, Revised 1997.

Janice VanCleave's A+ Science Fair Projects; Janice VanCleave, Jossey-Bass; 2003.

Science Fair Projects: Chemistry; Bob L. Bonnet; Sterling; 1999.

Scientific American Book of Great Science Fair Projects; Rosner, Marc; Jossey-Bass; 2000.

See for Yourself: More than 100 Experiments for Science Fairs and Projects; Vicki Cobb; Scholastic; 2001.

Strategies for Winning Science Fair Projects; Henderson, Tomasello; Jossey-Bass; 2001.

Successful Lab Reports: A Manual for Science Students; Lobban, Christopher; Schefter, Marla; 1992.

Additional Reference Books

Handy Science Answer Book, The, James Bobick and Naomi Baloba, Visible Ink. Press, 2002.

How to Think Like a Scientist Answering Questions by the Scientific Method, Stephen P. Kramer, Thomas Crowell, 1987. (Excellent for K-6.)

Young Person's Guide to Science Ideas That Change the World, A, Roy A. Gallant, Macmillan, 1993.

Science Supplies and Information

Carolina Biological Supply Co.
2700 York Rd., Burlington, NC 27215-3398
1-800-334-5551
www.carolina.com

Delta Education Inc., 80 Northwest Boulevard P.O. Box 3000
Nashua, NH 03061, 1-800-258-1302
www.delta-education.com

Home Training Tools, 1-800-860-6272.
www.hometrainingtools.com

Nasco Science, 901 Janesville Ave.,
Fort Atkinson, WI 53538, 1-800-558-9595.
www.enasco.com/science

Science Project Store, 972-470-0395
www.scienceprojects.net

Federal Agencies

From time to time website links do not work. Use the names of the organizations below to help find what you need utilizing a search engine. Many more agencies may be found on the ISEF website www.sciserv.org

Animals:

American Dairy Science Association
1111 N. Dunlap Ave., Savoy, IL 61874
(217) 356-5146
www.adsa.org

Animal and Plant Health Inspection Service
U.S. Dept. of Agriculture
1400 Independence Ave. SW, Washington, DC 20250
www.aphis.usda.gov/

Animal Welfare Information Center
National Agriculture Library
4th Floor, 10301 Baltimore Ave, Beltsville, MD 20705-2351
(301) 504-6212
www.nal.usda.gov/awic

Department of the Interior
Fish and Wildlife Service: Division of Endangered Species
1849 C. Street NW, Washington, DC 20240
www.fws.gov

Guidelines to Using Animals (pamphlet)
National Academy Press
500 5th St. NW Lockbox 285, Washington, DC 20055
1-888-624-8373
www.nap.edu

Guide to the Care and Use of Laboratory Animals (pamphlet)
Office for Protecting Animals from Research Risks
National Institutes of Health
9000 Rockville Pike Building 31, Room 5B63, Bethesda, MD 20892
http://newton.nap.edu/html/labrats/

Institute for Laboratory Animal Research
NAS 347 2101 Constitution Ave. NW, Washington, DC 20418
(202) 334-2590
ILAR@nas.edu
http://dels.nas.edu/ilar_n/ilarhome

Division of Endangered Species
www.fws.gov/endangered

Department of the Interior
1849 C Street NW, Washington, DC 2024
www.doi.gov

Hazardous Substances or Devices:

Safety in Academic Chemistry Labs (pamphlet—first copy free)
American Chemical Society
Office of Society Services
1155 16th Street NW, Washington, DC 20036
1-800-227-5558
http://membership.acs.org/c/ccs/pub_3.htm

Material Safety Data sheets
www.ilpi.com/msds/index.html

Office of Health and Safety
Centers for Disease Control and Prevention
1600 Clifton Road Atlanta, GA 30333
1-800-311-3435
www.cdc.gov/od/ohs/biosfty/biosfty.htm

Office for Protection From Research Risks (OPRR)
National Institutes of Health
200 Independance Ave, SW Washington DC 20201
1-877-696-6775
http://www.hhs.gov/ohrp

Health:

Division of Human Subjects
3935 University Way NE Seattle, WA 98105
206-543-0098
To have documents faxed, call 206-543-9218
http://www.washington.edu/research/hsd/index.php

American Psychological Association
750 First Street NE, Washington, DC 20002
(202) 336-5500 or 1-800-374-2721
www.apa.org
Information for students: www.apa.org/science/infostu.html

Medical Research:

National Institutes of Health
National Library of Medicine
8600 Rockville Pike, Bethesda, MD 20894
1-888-FIND-NLM (346-3656)
www.nlm.nih.gov

Prescription Drugs:

The Drug Enforcement Administration
Information Services Section
2401 Jefferson Davis Hwy., Alexandria, VA 22301
www.usdoj.gov/dea
Contact appropriate state agencies concerning additional regulations.

Office of BioTechnology Activities
National Institutes of Health
6705 Rockledge Drive, Suite 750, MSC 7985, Bethesda, MD 20892-7985
(301) 496-9838
www4.od.nih.gov/oba/

The Bureau of Alcohol, Tobacco, and Firearms
Distilled Spirits and Tobacco Branch
Room 8290 650 Massachusetts Ave. NW Washington, DC 20226
www.atf.treas.gov
Firearms and explosives: 304-616-4590

Sources of Cultures:

ATCC: The Global Bioresearch center
Human and Nonhuman Vertebrate Animal Tissue
American Type Culture Collection
P.O. Box 1549 Manassas, VA 20108
1-800-638-6597
www.atcc.org

Radiation/Radioactive Substances (Lasers) Or Safety and Health

Radioisotopes and Radioactive Substances
John Hickey
U.S. Nuclear Regulatory Commission
Material Safety and Inspection Branch
11555 Rockville Pike, Rockville, MD 20852
1-800-368-5642
www.nrc.gov

U.S. Department of Labor
Occupational Safety and Health Administration (OSHA)
Publications Office
200 Constitution Avenue NW, Washington, DC 20210
1-800-321-OSHA(6742)
www.osha.gov

Photo Credit Carol Benson

Bibliography

Bochinski, Julianne Blair. *The Complete Handbook of Science Fair Projects.* John Wiley & Sons, 1996.

46th International Science & Engineering Fair Rules. Science Service Inc., 1994, 2002, 2006.

Intel STS Science Services Online: www.sciserv.org.

International Rules for Precollege Science Research: Guidelines for Science and Engineering Fairs/ 2002-2003, 2006-2007. Science Services: www.sciserv.org/isef.

Iritz, Maxine Haren. *Blue Ribbon Science Fair Projects.* TAB Books, 1991.

Tocci, Salvatore. *How to Do A Science Fair Project.* Reed Business Information, Inc, 1997.

Photo Credit Kathi Kensinger

Media Angels Curriculum

Media Angels® Inc. offers award winning curriculum with a Creation-science focus. Visit us online at: www.MediaAngels.com

For a catalog write Felice@MediaAngels.com

Teaching Science and Having Fun! by Felice Gerwitz
This handy teacher's reference includes tips on how to schedule, setting a scope and sequence for grades K-12, setting up a lab on a shoestring budget, choosing supplies and resources, and much more! $12.95.

Virtual Field Trips: An Online Study Guide! by Felice Gerwitz
Go on a field trip without leaving home! This book includes step-by-step guidelines, lesson plans, questions and activities. $18.95.

The Truth Seeker's Mystery Series: by Felice Gerwitz and her homeschooled daughter Christina! These novels deal with the evolution vs. Creation debate! Join Anna and Christian Murphy, two homeschooled teens who encounter mystery, action and adventure as they learn that only the truth will set you free! For ages 12 and up. $8.99 each.

Volume #1: ***The Missing Link: Found!*** *Volume #2:* ***Dinosaur Quest at Diamond Peak*** *Volume #3:* ***Keys to the Past: Unlocked***

Literature Study Guides: by Felice Gerwitz Companion study guides for each of the novels, study questions with answers, further research and much more. $6.50 each

Truth Seekers Literature Study Guide: The Missing Link: Found!
Truth Seekers Literature Study Guide: Dinosaur Quest
Truth Seekers Literature Study Guide: Keys to the Past

Creation Study Guides!

Written by two homeschool moms: scientist and oil-well geologist Jill Whitlock, and Felice Gerwitz All are written on three levels: K-3, 4-8 & 9-12. They include vocabulary, experiments, resources and activities for each subject! $18.95 each

Creation Science: A Study Guide to Creation!
Creation Geology: A Study Guide to Fossils, Formations and the Flood!
Creation Astronomy: A Study Guide to the Gospel in the Stars!
Creation Anatomy: A Study Guide to the Miracles of the Body!

Activity Packs: These hands-on, reproducible sheets accompany the study guides above. All contain ready-to-use puzzles, games, activities and experiments. Use alone or with the unit studies. $12.95 each. Choose from ***Creation Science and Geology, Astronomy or Anatomy.***

About the Author

Felice Gerwitz, a mother of five chldren, has been homeschooling since 1986. She has degrees in elementary education, learning disabilities and early childhood education. Felice developed a love of science when she began homeschooling her children. One daughter and one son have both won several awards in science fairs, including upper-level awards. She has judged science fairs, held science workshops for children of all grade levels, and seminars for parents covering many topics. Felice, an award winning author, has written 16 titles to date. She continues to school her youngest children and lives in Fort Myers, Florida.

Newest addition grandaughter Emma Marie and son-in-law, William Moss

About the Publishers

Media Angels, Inc., is owned by Jeff and Felice Gerwitz. It was formed to publish high-quality materials with the busy family in mind. The Media Angels award winning curriculum is featured on many websites. For more information or a catalog, visit the website at www.MediaAngels.com or email Felice@MediaAngels.com.

Special Thanks

The author would like to thank the following people: Dr. Jay Wile for taking the time to read this book and for his advice; Carol Benson, Kathi Kensinger, Jackie Perseghetti and Veronique Scanlan, for permission to use their photos; Erica and Emily Benson, Caroline Kensinger, Christopher Scanlan, Ben Perseghetti and Nicholas and Anne Gerwitz for sharing their projects; and thanks especially to her daughter, Christina Moss for the hours spent updating the revised book.